Decoding The Universe 2

Decoding The Universe
2

Ian Beardsley

iUniverse, Inc.

New York Lincoln Shanghai

Decoding The Universe 2

iUniverse books may be ordered through booksellers or by contacting:

iUniverse
2021 Pine Lake Road, Suite 100
Lincoln, NE 68512
www.iuniverse.com
1-800-Authors (1-800-288-4677)

ISBN-13: 978-0-595-36540-1 (pbk)
ISBN-13: 978-0-595-80971-4 (ebk)
ISBN-10: 0-595-36540-X (pbk)
ISBN-10: 0-595-80971-5 (ebk)

Printed in the United States of America

Part 0 (introduction)

Humans settled in communities to farm and ranch, after hunting and gathering, perhaps 12,000 years ago, or more. Since that time, and long before it, the sun has been burning at the same temperature, more or less, the climate of the Earth varying in the Northern and Southern Hemispheres due mostly to their inclinations towards and away from the sun which alternate as the Earth goes around the sun, yearly (the earth is inclined to its orbit by about 23.5 degrees.) There are other factors that effect climate on earth, like ocean currents. The distances of the planets from the sun have been constant over geologic time, or more, as they orbit the sun. The elements are the same from sample to sample on earth, for all practical purposes. That is a gram of magnesium, for instance, weighs the same whether found in Alaska or Africa. The masses of the planets are constant. We mostly compare the molar masses of the elements (a mole being 6.02E23 atoms), and find where these ratios compare to the ratios formed between

1. Solar luminosity over a year and planetary kinetic energy (energy of motion around the sun).

2. Planetary masses

3. Separations between planets and the sun

We consider history, as it has been forged by the elements like copper for tools and much later for electrical wire. We are considering such relationships while the Titius-Bode Rule holds, since longer than the advent of man, which is an equation that predicts the distribution of the planets about the sun. It is geometric in nature.

The numbers for atoms, or chemical elements and compounds, are relative masses, which vary because the different kinds of matter are made up of units that have varying numbers of identical particles. These identical particles are three: protons, neutrons, and electrons. Protons and neutrons are almost the same mass, electrons are lighter. Neutrons have no charge, but protons are positively charged and electrons are negatively charged. Is what we mean by charged is that the particle has associated with it an electric field, which means that it can attract or repel particles that are charged. We say that opposite charges attract, and like charges repel. The number of particles that combine to make an atom, or element, determines its characteristics and quality. Elements are atoms, that cannot be reduced further unless through a so-called nuclear reaction. Compounds are combinations of atoms, and can be reduced to atoms through a chemical reaction. Nuclear reactions require allot of energy, as in what is occurring at the center of the sun, and stars. Chemical reactions can occur here on earth, in the laboratory, by mere mixture of the right elements, or by applying a flame. Most matter is neutral because it has an equal number of protons and electrons. If it has varying numbers of neutrons, we say that an element with more is simply a heavier version of the same. Elements that vary in weight alone are called isotopes. If an atom, or element, has an unequal number of protons and electrons, then it has a net charge and we call it an ion. More electrons and it is a negative ion, and less it is a positive ion. Electricity is the flow of electrons through a wire that is often an element such as copper or aluminum. Silver is the most conductive metal. Aside from being a metal, elements can be semimetals, or non-metals. Matter can exist in three states, solid, gas, or liquid. Some elements at earth temperatures are gases, like those that comprise the atmosphere. Though an element like Iron is a solid at earth temperatures, we can heat it in a forge and it becomes a liquid. It becomes a liquid when heated because adding energy makes its atoms further apart, and thus more movable. When we speak of particles, we are speaking of sub-atomic particles. Sub-atomic particles are actually composed of even smaller particles called quarks. We can use a series of magnets to accelerate sub-atomic particles to velocities that are a significant percentage of light velocity, and smash them into targets, to study how matter interacts. Photons are massless particles of light. Light can behave like an electromagnetic wave, or like a particle, depending on what we do with it. This is called wave-particle duality.

We are, in effect, trying to decode the universe, connect our history to its physical parameters, and find out in this metaphysical tablet of writing, why and how we got here, and it is looking like we can eliminate pure chance, though we are more scientific than to assume that means a supernatural being, like what we call god, had to have made the universe for us to arrive on the scene and sustain us, while on the other hand we cannot rule anything out so far.

The elements themselves were made by stars the heavier elements having been made by combining with the lighter elements under the pressure of the star by its own gravity. Energy is released in this nuclear reaction, a process called "fusion", and it is the outward force from this that causes a balance for the star with the inward force.

Part 1

Again, in this project, we compare the various relative masses of single atoms of the elements, and find where such quantities are more or less equal to the ratios between the masses of planets, their kinetic energies, and surface gravities, or even the energy emitted by the sun over the revolution of the earth about it once (a year). It is the chemical elements, which forged history, like iron and copper, and it is quite telling how they are related to the parameters of the solar system. The accuracy in determination of isotopic abundances including experimental error and variances in the natural occurrences of samples on earth make the determined values accurate to 99.9%.

It was Bob Dylan who sang: "You can't win with a losin hand". The question arises, have we been dealt a losing hand, or a winning hand? I would say, looking at the elements and what we are able to do with them, we have probably been dealt a winning hand, but we may be pushing the envelope in playing it out to the best interest of humanity. I think we should take a good look at what we need to do, democracy, environmental protection and human rights, and equality for all, freedom of speech and expression and do it.

Part 2

I believe that the two most important things to the human situation are warp drive and atomic, mind you not chemical, but atomic engineering. The reasons: warp drive because the distances between stars are so great, while the earth cannot take care of us forever, and atomic engineering because it would allow us to make from any element or compound any set of elements or compounds desired. Like, literally a taco or enchilada from dirt. The question I ask is: is it in the stars for us to do this?

Well, Mars is inhospitable enough. Its crust is basically silicon and its atmosphere carbon dioxide. Let us consider these.

$$CO_2 = 12.01 + 2(16.00) = 44.01$$

$$Si = 28.09$$

$$44.01/28.09 = 1.5667$$

Now consider what is any food but carbohydrates and/or proteins? Common to all carbohydrates are C, H, and O.

Common to all proteins are C, H, O and N.

$$(C+H+O)/N=(12.01+1.008+16.00)/14.01=2.07066$$

It just so happens that the two required minerals for the human diet, calcium and phosphorus, yeild:

$$Ca/P=40.08/30.97=1.29$$

This means that

$$[(CO_2)/(Si)](Ca/P)=(C+H+O)/(N)$$

Part 4

Iron was not separated from its ore until about 3500 B.C. in the Middle East, but around 4,000 B.C. iron was obtained from meteorites, rocks that came to earth from space, and used to make spear points and ornaments, in Sumeria and Egypt. That would have been 6,000 years ago from today, more or less, 2005, this the beginning of the Space Age (the first private venture ship leaves Earth atmosphere on June 21 2004, called SpaceShipOne). We now calculate how many miles the earth has traveled over that amount of time in its roughly 365-day journey around the sun.

The Earth orbit being nearly circular, its eccentricity varying over time from 0.0 to 0.06, it is sufficient to use $c=2(pi)r$ to calculate its circumference. The average distance of the Earth from the sun is 1.495979E13cm.

(1.495979E13cm)(m/100cm)(km/1000m)=1.495979E8km

(1.495979E8km)(3.14)(2)=9.39E8km

(9.39E8km/year)(6000years)=5.634E12km

The current separation between the closest star to the sun, Alpha Centauri, and the Sun, is 4.34 light years. One light year (ly) is 9.460530E17cm.

(9.46E17cm)(m/100cm)(km/1000m)=9.46E12km

5.634/9.46=0.59~0.60=3/5

Thus the distance the earth has traveled around the sun in the last 6,000 years, since iron was first crafted into spear points and so forth from meteorites, compared to the current distance between the closest star and the sun is nearly 3 to 5. Three and five are two consecutive Fibonacci numbers. The ratios between adjacent terms in the Fibonacci sequence are approximations to the golden number, or phi as it is called, which come closer and closer to the value as the numbers increase in value. Three and five are the fifth and sixth terms respectively.

I am no historian, and I have no idea of how accurate the date I have mentioned for these iron spear points and ornaments can be, but let us reverse the concept here and say that iron spear points were fashioned from meteorites 6,000 years ago because in that time from the beginning of the Space Age the Earth has traveled a fraction of the current distance to the nearest star from it of the golden number in its journey around the sun. In other words there is interconnectivity between all things, between events on earth that are human, even animal or plant in general, and what is going on through out the cosmos. Whether or not the first spear points were made from meteorites 6,000 years ago, it can be said that it was the beginning of the Egyptian calendar (4241 BCE the Egyptian calendar was established). What exactly was going on after the Earth had traveled around the sun exactly 0.618, the golden number, of the current distance to the nearest star, after this date? Keep in mind the separation between the solar system and this star changes ever so little over much time. Here it is calculated.

$$(0.618)(9.46E12km)=5.8E12km$$

$$(9.39E8km/year)(x)=5.8E12km$$

$$x=6,177 \text{ years}$$

$$6,177 \text{ years}-4,241 \text{ years}=1936 \text{ A.D.}$$

Around this time Alan Turing published his paper that founded the field of artificial intelligence (1937), and Theodosius Dobzhansky explained how evolution works (1937). The separation we give between Alpha Centauri and the Earth for 2005 was more or less the same in 1936. Nonetheless, we pointed out that 2004 was the first launching of SpaceShipOne.

Part 5

Iron and Bronze both have associated with them the word "age" for the meaning of "they were used over an epoch", we are still in the Iron Age. Having these in common their differences are that one is an element in its raw form, the other, bronze is an alloy, or is the combination of two elements. There is another common alloy, brass, which is an alloy of copper and zinc. Bronze is an alloy of the same copper, and tin. The Bronze Age was between the Stone Age and the Iron Age, beginning about 3500 B.C. using a bronze that was an alloy of copper and arsenic, arsenic gradually being replaced by the zinc. The Iron Age is marked by the wide spread use of iron from about 1500 B.C. to 1000 B.C. Brass was first used a great deal by the Romans 2,000 years ago. I write this in 2005 A.D. Copper was probably used as early as 8,000 B.C. because, existing in its raw form, it was easy to pound into tools, weapons and ornaments. No one knows when silver and gold were first used, but gold jewelry was unearthed in Bulgaria that dates back to about 4,000 B.C. on the coast of the Black Sea (Varna).

Copper, tin, and aluminum, all elements, they carry the majority of electrical currents for the majority of our electrical circuitry, always has and it's 2005 A.D.
Most ceremonial jewelry is either silver or gold, these elements actually run a current like no other metal, they are just too expensive, because of their rarity, to use a lot.
Brass is used for musical instruments, and iron and bronze and the same brass for weapons and tools. We would not have been able to fashion any of these if it were not for the heat we can generate from coal or coke, the latter a derivative of the former, and these basically carbon.
Before, of course, metals were used for farming or weapons stone was crafted into tools. The Stone Age, which was world wide, began about three and half million years ago.
Since three and a half million years ago, man's destiny has been determined by the substances, which are either the elements which were made in the stars, or their compounds. Spain set sail to the new world in fleets after looking for a westerly route to India. They discovered America. They returned in

9

great numbers because of the gold. Now gold may represent our currency, but it is worthwhile not just because it is a rare metal, but because it conducts electricity at extreme temperatures and is useful for covering our Space Shuttle so it can withstand extreme temperatures due to heat produced by friction from traveling at such high speeds through the atmosphere on re-entry. It can be pounded out into flat sheets. But we are entering a time of when he who needs gold can make it from anything including beach sand, for the elementary particles bound can be split apart by fission and reorganized into any combination of numbers by fusion. We are entering a time when we will be able to make silver from water the same way. Silver is the most efficient conductor of an electric current.

Part 6

When we burn anything (produce energy), it would seem to me we need oxygen to do it, or air, which I do admit has more nitrogen in it, but I think it is the oxygen that is used. Rockets burn hydrogen gas in oxygen, life burns stuff like sugars in oxygen, and coal is burnt in oxygen to smelt iron. Consider the bellows, it is an extension of the lung. The 2004 World book writes of it:

"Bellows is a device that produces wind by sucking air through one or more valves and then pumping it out. A simple bellows has a single air chamber, formed by two boards and soft leather sides. This bellows expels strong puffs of air through a nozzle. A hand-pumped simple bellows is commonly used to make wood in a fireplace burn intensely. A double bellows has two chambers and produces a continuous flow of air. More complex bellows help produce sound in such musical instruments as accordions and pipe organs."

Part 7

That substance we call air, which is the earth atmosphere where we breathe, is 21% oxygen gas and 78% nitrogen gas. O=16.00 and N=14.01, and since each are diatomic and the remaining one percent of the air is negligible, then the relative mass of air is:

2[16.00(.21)+14.01(.78)]=28.5756

The chemical formula for water is H_2O. Its mass is therefore

1.008(2)+16.00=18.016

if you refer to a periodic table of the elements. Let M_a represent the mass of air, and M_w represent the mass of water. Then

$(M_a)/(M_w)=28.5756/18.016=1.586$

This is close to the golden ratio, 1.618 and is close to its approximation by the ratio between the fifth and fourth terms of the Fibonacci sequence, 3 and 2 (3/2=1.5). But, most interestingly it is close to the ratio between the distance of Mars from the sun and the Earth from the sun. The average distance between Mars and the sun is
1.523 astronomical units. An astronomical unit is the average distance of the Earth from the sun, a distance that varies very little to begin with because the Earth orbits the sun to make nearly a perfect circle. So, if the distance to the Earth from the sun is r_e and the distance of mars from the sun is r_m, then

$(M_a)/(M_w) \sim (r_m)/(r_e)$

It is safe to say, I think, that air and water are crucial to most life, that they are at the heart of human necessity, and Mars and Earth are the only two planets capable of being inhabited by most life, definitely human life (Mars after it is terraformed). Mercury and Venus are too hot because of their proximity to the sun. The rest are gas giants except for Pluto, but there is debate as to whether or not Pluto should be called a planet as it is so small it might be best to call it a planetoid.

This last equation says something to the effect that the angular momentum of the solar system over the primordial disc that gave rise to it sorted out as air and water, much the way the heavier elements of a mixture go to the bottom of a bucket swung in a circle by a rope.

Now things get interesting.

Mass of earth=M_e=1.0000
Mass of mars=M_m=0.1
Mass of sugar(glucose)=C6H12O6=
(12.01)6+(1.008)12+(16.00)6=180.15
Mass of water=H2O=(1.008)2+16.00=18.016
C6H12O6/H2O~10

Incredible. This means that since

$$(M_e)/(M_m)=C6H12O6/H20$$

that

$$[M_e(water)]/[(M_m)(sugar)]=[M_a(r_e)]/[M_w(r_m)]$$

The square on water is the product of aluminum and carbon. That is

$$(water)^2=324.576\sim(Al)(C)=(12.01)(26.98)=324.0298$$

Therefore our last equation becomes

$$[(M_e)(Al)]/[(M_m)(sugar)]=[(M_a)(r_e)]/[C(r_m)]$$

Part 8

Plant life is quite a crucial component of human existence. Not only does it convert energy from the sun in the form of light to something we can burn, it makes oxygen gas (O_2) from the carbon dioxide (CO_2) that we exhale. Indeed it made the oxygen gas we now have in the earth atmosphere which was originally carbon dioxide. Mars is currently akin to the early earth atmosphere in that it is predominantly carbon dioxide. Thus, the problem currently arises how do we convert the Martian atmosphere into one of oxygen gas so we can inhabit that planet. Furthermore can we make oxygen in a spaceship using the CO_2 that we fill it with when we breathe in it? The answer is yes, we can us soda lime, but it would be more efficient if we could use something that is one compound. It is believed that potassium tetraoxide would be best, but we have yet to make it. Let us consider the components of potassium tetraoxide, K and O_4, and mars and the earth—for the above reasons—and see how they are connected.

Potassium (K)=39.10
O_4=4(16.00)=64
(64)/(39.10)=1.6368

If r_m=average Martian orbital distance and r_e=average earth orbital distance, then

$1.6 \sim (r_m)/(r_e) = 1.52$

$1.52/1.6 = .95$

an accuracy of 95%

Thus we say

$(O_4)/(K) \sim (r_m)/(r_e)$

This may be an indication that we can use potassium tetraoxide to make an earth-like habitat in a domed colony on Mars.

Part 9

Is the solar system an encoded tablet holding clues as to what our fate can be? I don't know, but I have not taken the path to a very high depth of education in the sciences and as such I have searched for a back door, and found allot of interesting relationships regarding nature. Since we possibly face a problem with global warming, what does this method say so far, if anything. Let's apply the method we have developed so far. Catastrophic climate change could theoretically be triggered if the so-called Atlantic Pump fails. This is where cold water that absorbs the heat retaining gas carbon dioxide (CO_2) fails to sink to the bottom of the ocean floor. This causes global warming because it is that cold water that absorbs the gas better than warmer water. The failure occurs (shutting down of the pump) because for some reason such as decreasing its salinity, which makes it heavy enough to sink, occurs because of the introduction of more water from melting polar ice caps and excess annual rainfall. So here I apply the method.

We start with the idea that Venus is a failed planet, once like the Earth it experienced a runaway greenhouse effect. The average distance of the Earth from the sun compared to that of the Venus is given by

$$(r\_e)/(r\_v)=1.39$$

where r_e is the average orbital distance of the Earth from the sun and r_v is that of Venus. The mechanism we spoke of concerning the Atlantic Pump is a form of one of the four cycles of nature, Carbon, Phosphorus, Water, and Oxygen are always made available to life after used. Since in global warming CO_2 is associated with water, its salinity, we compare the relative masses of these substances at their most fundamental level, these two of our four cycles:

$$(CO\_2)/(H\_2O)=44.01/18.016=2.44$$

This leaves phosphorus and oxygen:

$$(P)/(O)= 30.97/16.00=1.93$$

If

2.44=1.93x

then x=1.26

1.26/1.39=.91

Thus

$(CO_2)/(H_2O)=[(r_e)/(r_v)][(P)/(O)]$

within 91%.

This is where I always get stuck, I don't know what to do next.

Part 10

I would just like to say that in this divination where I converted from cubic kilometers to cubic centimeters, I called them liters which are actually milliliters (for the volume of the earth). In the end, however, the calculation works because I form a ratio between two numbers that are both off by the same factor.

Here is a sort of divination I have created that makes use of the interesting fact that the elements, copper, nickel, and so forth, have the same molar masses no matter where on earth they are found. Molar masses, the masses of the atoms, can vary throughout the universe, but those on earth must have a common origin. It is these masses that are responsible for their properties and appearances. To get on with it, I am connecting the molar masses of the elements as they occur on Earth, which is at a micro-level, a property of atoms, to things on a cosmic scale, but not beyond the solar system, as beyond it is where data starts to become inaccurate. For example where we don't know the actual size of the andromeda galaxy, we know with a great deal of certainty the distance to mars from the sun. Here is an example:

We calculate the mass of the quantity of air that occupies the volume of the Earth at the normal human body temperature, 37 degree C, and one atmosphere of pressure, which is the pressure at sea level. Then, we do the same thing for water, but at its freezing temperature which is 0 degrees C, a temperature at which water cannot be a gas, but we apply the gas law in any case, to be surreal, because there are no doubt universes where the laws are different (or so it is theorized, that they are uncountable in number). The result is interesting.

The radius of the Earth at its equator is 6,378 km. Its volume is approximately given by:

$V=(4/3)(pi)r^3$

Where r is its radius, which equals 6378km.

Therefore $r^3=(6378km)^3=2.59E11km^3$

$(2.59E11km^3)(pi)(4/3)=1.08E12km^3$

We now convert to liters.

$(1.08E12km^3)(1000m)^3(100cm)^3=1.08E27L$

Since the atmosphere is more or less 21% oxygen gas (O_2) and 78% nitrogen gas (N_2)

Air=2[(16.00)(0.21)+(14.01)(0.78)]=28.5756

We use the ideal gas law to determine n, the number of moles

$PV=nRT$

R=0.0821(atm)(L)/(mol)(K)

P=pressure(atm) T=temperature(K) V=volume(L) n=moles(mol)

The pressure at sea level is one atmosphere (atm).

We convert centigrade to Kelvin:

37 deg C + 273 deg = 310 deg K

[(1)(1.08E27L)]/[(0.0821)(310K)]=4.24E25mol

We now convert the moles to the mass:

(4.24E25mol)(28.5756g)/(mol)=1.21E27g of air

Now for water:

H_2O=(1.008)2+16.00=18.016

Converting to Kelvin:

(O deg C) + (273 deg) = 273 deg K

[(1atm)(1.08E27L)]/[(0.0821)(273K)]=4.8E25mol

Converting moles to the mass

(4.8E25mol)(18.016g)/(mol)=8.6E26g of water

We now divide grams of air by that of water:

1.21E27/8.6E26=1.4

This is equal to the division of gold (Au) by silver (Ag) multiplied by a factor of the life elements carbon over oxygen:

(Au/Ag)(C/O)=(197.0/107.9)(12.01/16.01)=1.4

We see that in this imaginary universe we have significance in relationship to the precious metals for which jewelry is made where the sacred volume of this life-bearing planet is concerned.

Even more accurately is what this says is that the mass of one atom of gold is to the mass of one atom of silver as the mass of air is to the mass of water at the normal human body temperature and freezing temperature of water, respectively.

Air and water are essential to human life, indeed to all life on Earth as far as I know and gold and silver are the precious metals used to make fine jewelry.

I would say this means that gold and silver are what I would call the life elements of the Earth.

My derivation says that silver and gold should be associated with the Earth where life is concerned, but according to astrology.com, silver is the moon element and gold is the sun element. I find everything ties together if you consider that the earth-moon distance (EM) is to the solar radius (SR) as silver (Ag) is to gold (Au). Here is the calculation:

EM/SR=3.84E10cm/6.9599E10=0.55

Ag/Au=107.9/197.0=0.55

That is the Earth is like a hinge to the moon and the sun, or pivotal, in other words, to silver and gold.

Gold when associated with air is equated with silver when associated with water and they in turn are associated with the sun and moon respectively where the Earth is concerned. So what is the next logical thing to do? I would say find out the meaning of Gold and silver in dreams. In dreams gold means "easy come, easy go" and silver means "ignoring spiritual in favor of materialism." What on earth do these have to do with air and water respectively. Well the air we breathe is free and and water cleanses.

Empedocles in his Tetrasomia associated the greek elements, earth, air, fire and water with the greek gods and goddesses Hera, Zeus, Hades, and Nestis (believed to be Persephone) respectively. But the Four Elements are also associated with the tetrahedron, earth, the hexahedron (cube), air the octahedron, water the icosohedron and dodecahedron the ether, and sphere the void. So lets extract from these the air and water as they are related to gold and silver:
Air: Zeus, octahedron
Water: Nestis, icosohedron
Now the octahedron and icosohedron are according to Buckminster Fuller two of the three structural systems, the third being the tetrahedron. That is, they are the only stable non collapsing flexcorner configurations. So, it looks like allot of doors start to open up.
It is said in astrology there are four elements, earth, air, fire, water, a square has its base the same as its height, and 3 modalities, for every two propositions there is an inference and the triangle is the minimal structure to enclose an area.
The product of 3 modalities and 4 elements is:
12 modality-elements
which is gold, since it is said in astrology gold embodies the properties of all of the metals.

Also 12 is divisible by more of its predecessors for its size than any other number (by 1,2,3,4,6).

I would let silver be represented by 9 since it is the next largest number divisible by 3 modalities, and it is the next most precious metal. We have

(3 modalities)(3modalities)=9modalites^2

So that

gold/silver=(12modality-elements)/(9modalities^2)=

(4 elements)/(3 modalities)

And therefore gold is the four elements and silver is the three modalities.

Also 4 is the 4+3=7 day weeks for the moon's complete cycle around the earth.

Part 11

Since other stellar systems may not even exist as we need them, and the distances between them are so immense, it might be better to unlock the mysteries of making them, and find the structure in ours that allows for so much life. There is, I have found, a correlation between the microworld and the macroworld, where our solar system is concerned. It may be related to why it is life bearing.

An interesting family of substances is methane (CH_4), ammonia (NH_3) and water vapor (H_2O). Methane is tetrahedral in structure, a carbon atom surrounded by 4 hydrogens. Ammonia is trigonal pyramidal, a nitrogen atom surrounded by 3 hydrogen atoms, and water vapor is triangular, or bent, an oxygen atom surrounded by two hydrogens. These represent stable structural systems as they are all systems of triangles, which are the only stable polygons. These substances combined under energy with hydrogen gas form amino acids, the building blocks of life. The core atoms of these molecules, carbon, nitrogen, and oxygen, are all in period two of the periodic table and follow directly one after the other, and are all in amino acids, the hydrogen as well. It is a hypothesis of astrobiology that amino acids formed in the protoplanetary cloud before the earth ever formed. In this sense we may have our origins in deep space. Is what I mean by structural systems is that there are only three structural systems, the tetrahedron, the octahedron, and the icosohedron. They are the only stable solids, that is noncollapsing flex corners whose faces are triangles. Most compounds are something other than these, like pentagons with linear off shoots for example, that comprise the wrong number of atoms to make a "solid" unit, and I mean solid as in the pythagorean solids, the geometric term. Both methane and ammonia make different variations on the tetrahedron, a pythagorean solid.

When plants perform photosynthesis, they combine carbon dioxide with water and release oxygen. The reaction is:

$$CO_2 + 2H_2O \longrightarrow^{\#} CH_2O + O_2 + H_2O$$

As can be seen a sugar is made. Important to most plants to do this is Nitrogen. Nitrogen (N_2) is the most abundant gas in the earth atmosphere, comprising about 78.03% of it. We now calculate the molecular masses of these special gases:

$CH\_4=(12.01+4(1.01))=16.05$
$NH\_3=(14.01+3(1.01))=17.04$
$CO\_2=(12.01+2(16.00))=44.01$
$H\_2O=(2(1.01)+16.00)=18.02$
$N\_2=(14.01+14.01)=28.02$
$O\_2=(16.00+16.00)=32.00$
We now form some ratios between these molecular masses:
$(O\_2)/(CH\_4)=32.00/16.05=1.992\sim 2$
$(NH\_3)/(CH\_4)=17.04/16.06=1.061\sim 1$
$(CO\_2)/(O\_2)=44.01/32.00\sim 1.4=sqrt(2)$
$(CO\_2)/(N\_2)=44.01/28.02\sim 1.6=(sqrt(5)+1)/2=phi$
$(O\_2)/(H\_2O)=32.00/18.02=1.776\sim sqrt(3)$
Notice that these values are given by the sequence:
$|2cos(pi/n)|$ n=(1,2,3,4,5,6)(pi/n)radians
Observe:
$2=|2cos(pi)|$
$0=|2cos(pi/2)|$
$1=|2cos(pi/3)|$
$sqrt(2)=|2cos(pi/4)|$
$(sqrt(5)+1)/2=phi=|2cos(pi/5)|$
$sqrt(3)=|2cos(pi/6)|$
Geometrically sqrt(2) is the ratio of the side of a square to its radius. Phi is the ratio of the chord of a regular pentagon to its side. Sqrt(3) is the ratio of the side of an equilateral triangle to its radius, and 1 is the ratio of the side of a regular hexagon to its radius. The square, the regular hexagon and the equilateral triangle are the tessellating regular polygons.

Part 12

We compare the mass of the earth to the mass of the sun, and multiply that ratio by the distance between them. Let the mass of the earth be M_e, and the mass of the sun be M_s. Let the distance between them be r.

(M_e/M_s)r=(5.976E27/1.989E33)(1.495979E13)=(4.495E7)cm=449.5km
We now divide that result by the radius of the earth, R_e:

(449.5km)/(6378.5km)=0.07

Hydrogen is the most abundant element in the universe and Nitrogen is the most abundant element in the earth atmosphere. We now compare their molar masses:

(H/N)=(1.01)/(14.01)=0.07

And we see that

(H/N)=((M_e)(r))/((M_s)(R_e))

Having showed the last equation, where hydrogen is the most abundant element in the universe and nitrogen is the most abundant element in the earth atmosphere, then since Mars is a terrestrial planet upon which we can set foot as opposed to
Venus and Mercury, let's apply the same idea to mars. The most abundant gas in the Mars atmosphere is carbon dioxide, or CO_2. It in fact comprises 95.3% of its atmosphere. We have:

H/(CO_2)=1.01/44.01=0.02

Now let M_m= mass of mars, M_s = mass of the sun, r= the distance between them, and R_m= the radius of mars. We have

$(M\_m)(r)/(M\_s)(R\_m)=(6.418E26)(2.279409E13)/$
$(1.989E33)(3.393096E8)=0.02$

And therefore,

$H/(CO\_2)=(M\_m)(r)/(M\_s)(R\_m)$

Keep in mind these equations, both for the earth and mars, hold for a solar system at its peak as an orderly arrangement of parts. Eventually it will begin to degenerate. The sun is losing mass every day and therefore r, for any of the planets, will grow.

Thus we say in general:

$H/A=(M\_p)r/(M\_s)(R\_p)$
where H is the molar mass of hydrogen, A is the molar mass of the most abundant element or gas in the planet's atmosphere, (M_p) is the mass of the planet, (M_s) is the mass of the star, r is the distance between the planet and the star and (R_p) is the radius of the planet. Lets look at the quantity $(M\_p)r/(M\_s)$. It is equal to $(d\_1)/(d\_2)(d\_1+d\_2)$, the ratio of the distances between the balancing point of a cosmic teeter totter and the planet and the star balanced on it, times its length. We then compare such a distance to the radius of the planet.

Part 13

The relative equatorial surface gravities uncorrected for centrifugal force of the earth and mars respectively are 1.000 and 0.380. Their proportions are

1.000/0.380=2.63

The ratio of the molar mass of oxygen gas to that of carbon is

$(O_2)/C=32.00/12.01=2.66$

Thus, $(g_e)/(g_m)\sim(O_2)/C$

where g_e is the equatorial surface gravity of the earth and g_m is the equatorial surface gravity of mars. The centrifugal forces being nominal, this says it takes the same amount of energy to lift a mole of carbon on the earth as it does to lift a mole of oxygen gas on mars the same distance if the atmospheric pressures are excluded. Carbon is the basis of life and oxygen gas its necessity (for human life).

The data for this study came from the Handbook Of Space Astronomy And Astrophysics, by Martin V. Zombeck, Cambridge University Press, 1982.

Part 14

Luminosity of the sun=3.826E26J/s=L

seconds in year=3.1536E7s=t

Mean orbital velocity of earth=29790m/s=v

Mass of the earth=5.976E24kg=m

Lt=1.2E34J

$(1/2)mv^2$=2.65E33=kinetic energy of earth

$Lt/(1/2)mv^2$=4.5

Now consider the molar masses of iron and carbon:

Fe/C=55.85/12.01=4.6

Thus

Lt/(1/2)mv^2=Fe/C

Thus the comparison of the annual energy output of the sun in light, to the kinetic energy of the earth, or to its energy of motion in other words, is the same as the comparison of iron to carbon as far as the weight of an atom is of the former to the latter. Keep in mind a year represents one complete revolution of the earth about the sun and that the iron age represented a revolution in tool making while carbon is the basis of life. We have used the 365 day year for this calculation, and only considered the kinetic energy of the earth do to its orbital motion.

Part 15

It is natural now to ask what is the comparison of the density of Iron to that of carbon at earth temperatures and pressures:

Fe=7.87g/cm^3 and C=2.26g/cm^3

7.87/2.26=3.48=Ag/P=107.9/30.97

Ag is silver and P is phosphorus. Silver is the most conductive metal of heat and electricity and the most reflective, being used therefore in electronics and to make telescope mirrors. Phosphorus is necessary to animal and plant life and is returned to life after used in one of the four natural ecological cycles.

Washington Alloy Co writes:

"Washington Alloy Phos-Copper-Silver Brazing Alloys (USA 0, USA 2%, USA 5%, USA 6F and USA 15%) are all manufactured to offer economy as well as consistently high standards of quality and performance. These alloys are excellent for joining copper to copper where the phosphorus content of the phos-copper-silver brazing alloy reacts with the copper of the base metal in such a way that the filler metal becomes self-fluxing. For this reason these alloys are used quite extensively for joining closed copper tubing in the refrigeration and air conditioning industries where flux removal after brazing is difficult to impossible."

World Book 2004 writes:

"Copper is the best low-cost conductor of electric current. As a result, the electrical industry uses about 60 percent of the copper produced, chiefly in the form of wire. Copper wire carries most of the electric current inside homes, factories, and offices. Large amounts of copper wire are used in telephone systems, as well as in television sets, motors, and generators."

Thus the mechanics of the solar system have taken us to copper, among other things, all key to technology in their various ways. Copper occurs free in nature, but we get most of it from copper sulfides. We separate it from copper sulfide in the following reaction:

$Cu_2S(l)+O_2(g) —>2Cu(l)+SO_2(g)$

Thus the pollutant sulfur dioxide is released into the air, a major contributor to acid rain. If we relate Cu, SO_2 and S to Ag and P, we have come all the closer to finding the connection of human activity to the motion of the earth as related to the output of solar energy during this phase in human life where it is beginning to enter space.

We have

$(Cu_2)/S=3.96=A$ and $(SO_2)/Cu=5.33=B$

We also have that $Ag/P=3.48=C$ and

$(densityAg)/(densityP)=5.7692=D$

(densities at earth temperatures and pressures)

AB=21.1068 and CD=20.07

Thus,

AB=CD within 95%

Notice that $(O_2)/(S) \sim 1.00$

O_2 we breathe, but combined with S is poison.

If something is readily available in nature, odds are it is there for us to use, but the second we have to make it, extract from it from something, or alter it, we will upset the natural structure that maintains a healthy balance for life, ourselves a component of that structure. But perhaps components can change, if other components change in the right way, including ourselves (i.e. the story of evolution).

As you can see we are on a quest that is about to fullfill itself, as all of this leads us to coke and coal, I am not sayin how, but obviously it is a dead give away as its importance surmounts as we run out of oil while on the verge of entering a real space age, and that it is to energy production (electricity) and metalurgy perhaps what copper and silicon has been to electronics. It has also been used since no one knows when, but the chinese started its production as an industry in…

Thus we compare copper to silicon, and equate it to coal, C, which is carbon, compared to hydrogen and find that they are equal by a factor x, which, is the ratio of the mean jupiter-sun distance divided by the mean earth-sun distance:

$C/H = x(Cu/Si)$ x=5.2

Hydrogen is compared to coal, because it too is an energy source, the one that is proposed to take the place of fossil fuels, technological breakthrough pending to separate it from water so that more energy is provided than is put into the process. Doped silicon represented a revolution in electronics in the form of integrated circuitry. Jupiter should become our source of hydrogen gas. Hydrogen gas comprises 86% of its atmosphere. Copper and coal have been man's work horse in the past, silicon and hydrogen his future. As elements of the past were important they become important again, but in a different way. Thus carbon in the form of coal, heated our homes, and smelted Iron, now it shows promise in the potential of becoming a one dimensional quantum wire. Silver and gold await their place in the future, the vast reservoirs of jewelry converted into circuitry because of their special, and important electrical properties. Silicon is on the border between the recent past and the future. Mars is the one terrestrial planet we can colonize. Thus in the spirit of our last equation we write:

$$Ag/Al=y(Si/C)$$

where Ag is silver, Al is aluminum, Si is silicon and C is carbon. "y" is the ratio of the average mars-sun distance to the average earth-sun distance. y=1.52

The work horse of electronics is tin-lead solder (Sn and Pb) and the workhorse of chemistry is pyrex glass, basically silica (SiO_2) and boron (B). Let z=9.54 the ratio of the Saturn-sun distance to the earth-sun distance, and we have, in the spirit of our last two equations

$$(SiO_2)/B=z(Sn)/Pb$$

I think creating some categories is called for here:

Reservoirs: coal, hydrogen gas, methane etc...

Hardware: solder and solder iron, beakers, and test tubes, glassware etc...

Systems: silver wire, copper wire, aluminum wire etc...

Software: integrated circuits, transistors, diodes, resistors, capacitors, components.

We can now write by the last three equations:

Jupiter (meaning) reservoir/system/software=technology

Shelter: regular hexagon

Mars (meaning) software/system=magic

Food: square

Saturn (meaning) hardware=industry

Self: equilateral triangle

The dual of the regular hexagon is the equilateral triangle, the dual of the square is itself, and the dual of the equilateral triangle is the regular hexagon.

Part 16

The strategy: Invariances in nature are searched for so they can be used to form ratios that can be compared. A reason for the harmonies is sought. Intuition is utilized. (math notes):The idea is to find the unifying themes of each subject, and unify those under a common theme so as to make it easier to remember and more effective in its applications.

It is becoming increasingly clear that we should enter a space age, and, as such, only use coal and coke for metalurgy, gold for electronics and get gases for transportation from the gas giants, like Jupiter, Saturn, and Neptune (methane and hydrogen) and pipe in our electricity from states such as oregon which have a great surplus do to their immense river power. Remember, the clean burning fuel cell auto can run off of hydrogen.

Part 17

The formula for the legs of right triangle given its area and hypotenuse is:

a,b=+/-sqrt((h^2+/-sqrt(h^4-16A^2))/2)

Earth surface area is 5.1E8 km^2

Solar radius is 696,000 km

Using these values in our formula, we get

a=6.957E5 km and

b=2.0433E4km

a/b=34

Now

a/b=[(mass of neptune)O]/[(mass of earth)N]~30(16/14)

O=16.00=oxygen and N=14.01=nitrogen

Part 18

Given the golden ratio occurs throughout life, one might ask where it would occur in artificial intelligence. Now silicon, oddly enouph is in the same group as carbon. As we are carbon based it is interesting that artificial intelligence, or man made intelligence, would be silicon based. Silicon is pretty much useless, as I understand it, if it is not "doped" with boron or phosphorus, atleast as far as negative and positive type silicon are used. Let us consider phosophorus is for making negative type silicon and boron is for making positive type silicon. Let us find the geometric mean between phosphorus (P) and boron (B) and divide it by silicon (Si):

sqrt(P*B)/Si=sqrt(30.97*10.81)/(28.09)=0.65

and let us take the harmonic mean between phosphorus and boron and divide it by silicon:

(2*(30.97*10.81)/(30.97+10.81))/28.09 =0.57

Now let us take the arithmetic mean of these two numbers:

(0.65+0.57)/2=0.61

which are the first two digits in the golden ratio.

With the advent of plant life, the earth began to accumulate oxygen in its atmosphere. Before that, there was a lot of CO_2. Current levels are 21% oxygen, 78% nitrogen. I suggest that the ideal levels are 25% oxygen, 75% nitrogen so that

$$[(CO_2)/(O_2)][(\%N_2)/(\%N_2+\%O_2)] \sim [sqrt(2)](3/4) \sim 1.00$$

The Earth atmosphere is nearly 25% oxygen and 75% nitrogen. Oxygen nearly weighs the same as nitrogen. Both occur as diatomic molecules. O=16.00 and N=14.01~14.00.
Let us say imagine that there was such a thing as an air atom, then, its relative mass would be:

$$(16.00)(0.25)+(14.00)(0.75)=4+10.5=14.5=air.$$

If it was diatomic, then air2=29.00: somewhere between silicon and phosphorus, which are next to one another in the periodic table of the elements. Silicon is used in combination with phosphorus for making diodes, and ultimately transistors and integrated circuits.

Part 19

Project Genesis realized:

Luminosity of the sun (L) =3.826E26J/s

Mass of the sun (M) = 1.989E30Kg

The luminosity of a main sequence star like the sun is proportional to the mass by:

L is proportional to M^3.5

We introduce a constant of proportionality (a) and calculate it based on the sun, that is

L=aM^3.5

And we find a=0.25J/(Kg)^3.5(s) correction:a=3.8E-79

Now the exponent 3.5 equals Au/Fe=196.97/55.85 where Au is gold and Fe is iron.

And the constant of proportionality, a, equals Al/Ag=26.98/107.87=0.25 where Al is aluminum and Ag is silver.

Now, Iron is the best metal for tools (early agriculture, the iron age)

And Gold is the most conductive at extreme temperatures (the space age)

Aluminum is the most abundant metal in the earth crust (electronics)

And silver is the most electrically conductive metal (let us say computers)

Thus, our project genesis equation is:

L=(Al/Ag)M^(Au/Fe)

Al/Ag in J/(Kg^3.5)s

We have talked about the signifcance of the metric system where the constant of proportionality is concerned in "approaching the key".

Part 20

Magnesium (Mg) is the lightest metal for construction purposes, i.e. it is good for spacecraft construction in that it provides for a low mass to thrust ratio and boron (B) is neccessary for navigation systems in the construction of system computers in that it is used to make positive type silicon, and thus the ratio of Mg/B encodes these principles in that the only terrestrial planet that can be colonized is mars from earth and the the ratio is equivalent to the ratio of the earth's escape velocity to that of mars, the velocities needed to be attained to not fall back to these respective planets. These ratio's are nearly equivalent as well to oxygen gas (O_2) to N (nitrogen) which are the components of chemical aeronomy.

$(O_2)/N=32.00/14.01=2.28\sim(v_e)/(v_m)=11.2/5.0\sim Mg/B=2.25$

where v_e=escape velocity of earth and v_m that of mars.

Mg=24.31amu and B=10.81amu

Part 21

Plants are the link between humans and the sun, they convert energy from the sun into sugar, $C_6H_{12}O_6$. A gram of this, called glucose, can yield15.6kJ of energy when burned. The reaction is:

$C6H12O6+6O2—>6CO2+6H2O$

$(15.6E3J/g)(3E-22g/molecule)=$

$4.68E-18J/molecule$

molar mass of glucose:

$C6H12O6=6(12.01)+12(1.01)+6(16.00)=180.18amu$

$(180.18g/mol)(1mol/6.02E23molecules)=3E-22g$

Luminosity of sun $(L)=3.826E26$ J/s

$(365days)(24hrs)(60min)(60s)=3.15E7$ s/yr

$L(s/yr)=(3.826E26J/s)(3.15E7s/yr)=1.2E34$ J/yr

r=earth's distance from the sun=1.5E11m

R=sun radius=7E8m

LR^2/r^2=(1.2E34J/yr)(7E8m)^2/(1.5E11m)^2=2.6E29J/yr

This last calculation is energy per year in sun light at earth orbit where a year is the time it takes for one revolution of the earth about the sun.

(2.6E29J/yr)(molecules/4.68E-18J)=5.6E46molecules/yr

(5.6E46molecules/yr)(2.99E-22g/molecule)=1.67E25g/yr

This last figure is the amount of glucose that can be produced at earth orbit over one complete orbit of the earth around the sun (one year).

Earth mass=5.976E27g

5.976E27g/1.67E25g=357

M_j=mass of jupiter=318

M_e=mass of earth1.00

(M_j/M_e)~357(N/O)

where N=14.01, O=16.00

N is nitrogen and O is oxygen.

0.36=N/K

where N is nitrogen, the most abundant element in the earth atmosphere and is key to nitrogen fixation and K is potassium, key to photosynthesis.

(surface area of earth/2)=4(pi)(6.38E6m)^2=2.6E14m^2

Neccessary in making the sugar from sun light is chlorophyll, pigements in leaves that carry out photosynthesis. In order for plants to make chlorophyll, they need Potassium (K).

The way I did this calculation, may be important. For instance, when computing the number of seconds in a year, I used the 365 day year, which does not account for leap year every four years, where an extra day is added to the year.

Part 22

We have shown in "genesis itself" that the glucose made by a plant can produce

4.68E-18J/molecule of energy when burned. Hydrogen gas burned in oxygen in the reaction (rocket fuel):

$H_2+(1/2)O_2 —> H_2O$

can produce 120 kJ of energy per gram

$H_2=2.02$

(2.02g/mole)(1 mole/6.02E23atoms)=

$3.35E-24g/(moleculeH_2)$

$(3.35E-24)(120E3)=4.02E-19J/moleculeH_2$

(4.68E-18)/4.02E-19=11.64=W/O

where W=tungsten and O=oxygen

W=184 and O=16

Tungsten is used to make light (i.e. is the filament of a light bulb) and oxygen is used to burn fuel, whether it be a sugar or hydrogen gas.

Part 23

After crafting tools of stone, humans began to use bronze, an alloy of copper and tin. Iron was the first non alloy to be used for the same. So far we have considered carbon, oxygen and nitrogen. Here is the key to project genesis:

$(Cu+Sn)/Fe=2(C+N)/O$ and $H\sim1.00$

NaCl = table salt = 22.99+35.45=58.44 and Ni=58.69=nickel

Thus NaCl=Ni

Bronze=Cu+Sn and Brass=Cu+Zn

Bronze/Brass = 182.26/128.94=1.414=sqrt(2)

Cu=copper, Sn=tin, Zn=zinc

Cu=63.55, Sn=118.71, Zn=65.39

Fe=55.85

With the advent of plant life, the earth began to accumulate oxygen in its atmosphere. Before that, there was a lot of CO_2. Current levels are 21% oxygen, 78% nitrogen. I suggest that the ideal levels are 25% oxygen, 75% nitrogen so that

$[(CO_2)/(O_2)][(\%N_2)/(\%N_2+\%O_2)]\sim[sqrt(2)](3/4)\sim1.00$

Part 24

We have said that:

Jupiter (meaning) reservoir/system/software=technology
Shelter: regular hexagon
Mars (meaning) software/system=magic
Food: square
Saturn (meaning) hardware=industry
Self: equilateral triangle

The dual of the regular hexagon is the equilateral triangle, the dual of the square is itself, and the dual of the equilateral triangle is the regular hexagon.

Thus, to continue, technology gives birth to industry and industry to technology and magic gives birth to magic. Thus Mars stands alone while Jupiter and Saturn are coupled.

We can now say that Mars represents godliness, Jupiter creation and Saturn rebirth.

Mars is a sort of extension of Earth after modified—while perhaps Jupiter and Saturn are sources of hydrogen gas for rocket fuel for surface to orbit operations, necessary for building interplanetary mother ships in Mars and Earth orbit that are nuclear, preferably using the very same hydrogen for their propulsion as well. Probably after entering this phase, if we do, it will be when we will start making some breakthroughs in propulsion, hyperdrive, antigravity—that can carry us to the stars within a reasonable amount of time.

It may be that in a certain respect we can draw the following energy comparison:

(nuclear power)/(chemical power)=(coal)/(fire wood)=(hydro)/(wind)

(nuclear/chemical)=Jupiter

(coal/firewood)=mars
(hydro/wind)=saturn

In a sense nuclear power (H, or hydrogen) is connected to coal (C) and hydro-electric (H_2O) in the following reaction:

$$C(s)+H_2O(g)\#CO(g)+H_2$$

Which is used to convert coal into the clean burning H_2. Keep in mind it is the carbon in coal that is an energy source, and it is hydrogen that a main sequence star begins to fuse. CO, the poisonous gas carbon monoxide, is used to separate iron from its ore.

And as sugar burns fast, carbohydrates slower but longer, and proteins slowest but longest, we have
Carbohydrates correspond to Jupiter, proteins to mars, and sugars to Saturn.
I propose that the igloo represents Saturn, the tipi Jupiter and the yurt mars. As such we have related heat and light, food and habitats to these three key planets.
Question: thus eating meat in a yurt and burning coal in it to keep you warm somewhere in Mongolia has what to do with mars?

Part 25

Formulas Derived from the Parallelogram

Remarks. Squares and rectangles are parallelograms that have four sides the same length, or two sides the same length. We can determine area by measuring it either in unit triangles or unit squares. Both are fine because they both are equal sided, equal angled geometries that tessellate. With unit triangles, the areas of the regular polygons that tessellate have whole number areas. Unit squares are usually chosen to measure area.

Having chosen the unit square with which to measure area, we notice that the area of a rectangle is base times height because the rows determine the amount of columns and the columns determine the amount of rows. Thus for a rectangle we have:

$A = bh$

Drawing in the diagonal of a rectangle we create two right triangles, that by symmetry are congruent. Each right triangle therefore occupies half the area, and from the above formula we conclude that the area of a right triangle is one half base times height:

A=(1/2)bh

By drawing in the altitude of a triangle, we make two right triangles and applying the above formula we find that it holds for all triangles in general.

We draw a regular hexagon, or any regular polygon, and draw in all of its radii, thus breaking it up into congruent triangles. We draw in the apothem of each triangle, and using our formula for the area of triangles we find that its area is one half apothem times perimeter, where the perimeter is the sum of its sides:

A=(1/2)ap

A circle is a regular polygon with an infinite amount of infitesimal sides. If the sides of a regular polygon are increased indefinitely, the apothem becomes the radius of a circle, and the perimeter becomes the circumference of a circle. Replace a with r, the radius, and p with c, the circumference, and we have the formula for the area of a circle:

A=(1/2)rc

We define the ratio of the circumference of a circle to its diameter as pi. That is pi=c/D. Since the diameter is twice the radius, pi=c/2r. Therefore c=2(pi)r and the equation for the area of a circle becomes:

$A=(pi)r^2$

Math Notebook of Ian Beardsley

(More derived from the parallelogram)

Divide rectangles into four quadrants, and show that

A. $(x+a)(x+b)=(x^2)+(a+b)x+ab$

B. $(x+a)(x+a)=(x^2)+2ax+(a^2)$

A. Gives us a way to factor quadratic expressions.

B. Gives us a way to solve quadratic equations. (Notice that the last term is the square of one half the middle coefficient.)

Remember that a square is a special case of a rectangle.

There are four interesting squares to complete.

1) The area of a rectangle is 100. The length is equal to 5 more than the width multiplied by 3. Calculate the width and the length.

2) Solve the general expression for a quadratic equation, $a(x^2)+bx+c=0$

3) Find the golden ratio, a/b, such that $a/b=b/c$ and $a=b+c$.

4) The position of a particle is given by $x=vt+(1/2)at^2$. Find t.

Show that for a right triangle $(a^2)=(b^2)+(c^2)$ where a is the hypotenuse, b and c are legs. It can be done by inscribing a square in a square such that four right triangles are made.

Use the Pythagorean theorem to show that the equation of a circle centered at the origin is given by $r^2=x^2+y^2$ where r is the radius of the circle and x and y the orthogonal coordinates.

Derive the equation of a straight line: $y=mx+b$ by defining the slope of the line as the change in vertical distance per change in horizontal distance.

Math Notebook of Ian Beardsley

Triangles

All polygons can be broken up into triangles. Because of that we can use triangles to determine the area of any polygon.

Theorems Branch 1

1. If in a triangle a line is drawn parallel to the base, then the lines on both sides of the line are proportional.

2. From (1) we can prove that: If two triangles are mutually equiangular, they are similar.

3. From (2) we can prove that: If in a right triangle a perpendicular is drawn from the base to the right angle, then the two triangles on either side of the perpendicular, are similar to one another and to the whole.

4. From (3) we can prove the Pythagorean theorem.

Theorems Branch 2

1. Draw two intersecting lines and show that opposite angles are equal.

2. Draw two parallel lines with one intersecting both. Use the fact that opposite angles are equal to show that alternate interior angles are equal.

3. Inscribe a triangle in two parallel lines such that its base is part of one of the lines and the apex meets with the other. Use the fact that alternate interior angles are equal to show that the sum of the angles in a triangle are two right angles, or 180 degrees.

Theorems Branch 3

1. Any triangle can be solved given two sides and the included angle.

c^2=a^2+b^2-2abcos(C)

2. Given two angles and a side of a triangle, the other two sides can be found.

a/sin(A)=b/sin(B)=c/sin(C)

3.Given two sides and the included angle of a triangle you can find its area, K.

K=(1/2)bc(sin(A))

4.Given three sides of a triangle, the area can be found by using the formulas in (1) and (3).

Question: what do parallelograms and triangles have in common?

Answer: They can both be used to add vectors.

Math Notebook of Ian Beardsley

Trigonometry

When a line bisects another so as to form two equal angles on either side, the angles are called right angles. It is customary to divide a circle into 360 equal units called degrees, so that a right angle, one fourth of the way around a circle, is 90 degrees. The angle in radians is the intercepted arc of the circle, divided by its radius, from which we see that in the unit circle 360 degrees is 2(pi)radians, and we can relate degrees to radians as follows:

Degrees/180 degrees=Radians/pi radians

An angle is merely the measure of separation between two lines that meet at a point.

The trigonometric functions are defined as follows:

cos x=side adjacent/hypotenuse

sin x=side opposite/hypotenuse

tan x=side opposite/side adjacent

csc x=1/sin x

sec x=1/cos x

cot x=1/tan x

We consider the square and the triangle, and find with them we can determine the trigonometric function of some important angles.

Square (draw in the diagonal): cos 45 degrees =1/sqrt(2)=sqrt(2)/2

Equilateral triangle(draw in the altitude): cos 30 degrees=sqrt(3)/2; cos 60 degrees=1/2

Using the above formula for converting degrees to radians and vice versa:

30 degrees=(pi)/6 radians; 60 degrees=(pi)/3 radians.

Math Notebook of Ian Beardsley

The regular hexagon and pi

Tessellating equilateral triangles we find we can make a regular hexagon, which also tessellates. Making a regular hexagon like this we find two sides of an equilateral triangle make radii of the regular hexagon, and the remaining side of the equilateral triangle makes a side of the regular hexagon. All of the sides of an equilateral triangle being the same, we can conclude that the regular hexagon has its sides equal in length to its radii. If we inscribe a regular hexagon in a circle, we notice its perimeter is nearly the same as that of the circle, and its radius is the same as that of the circle. If we consider a unit regular hexagon, that is, one with side lengths of one, then its perimeter is six, and its radius is one. Its diameter is therefore two, and six divided by two is three. This is close to the value of pi, clearly, by looking at a regular hexagon inscribed in a circle.

The sum of the angles in a polygon

Draw a polygon. It need not be regular and can have any number of sides. Draw in the radii. The sum of the angles at the center is four right angles, or 360 degrees. The sum of the angles of all the triangles formed by the sides of the polygon and the radii taken together are the number of sides, n, of the polygon times two right angles, or 180 degrees. The sum of the angles of the polygon are that of the triangles minus the angles at its center, or A, the sum of the angles of the polygon equals n(180 degrees)-360 degrees, or

A=180 degrees(n-2)

With a rectangular coordinate system you need only two numbers to specify a point, but with a triangular coordinate system—three axes separated by 120 degrees—you need three. However, a triangular coordinates system makes use of only 3 directions, whereas a rectangular one makes use of 4.

A rectangular coordinate system is optimal in that it can specify a point in the plane with the fewest numbers, and a triangular coordinate system is optimal in that it can specify a point in the plane with the fewest directions for its axes. The rectangular coordinate system is determined by a square, and the triangular coordinate system by an equilateral triangle. They are the basis for many mosaics in Moorish castles, such as those in the Alhambra in Spain.

From the Physics Notebook of Ian Beardsley
F=ma M=mv v=x/t
F=Force M=momentum m=mass v=velocity x=distance t=time a=acceleration
M=mv=m(x/t) a=dv/dt a(dt)=dv v=at v=dx/dt dx/dt=at dx=at(dt) x=(1/2)at^2
x=x_0+vt+(1/2)at^2

int[x^n] 0 to x =$(x^{(n+1)})/(n+1)$ and $(d/dx)x^n = nx^{(n-1)}$

K=kinetic energy U=potential energy C=constant

K=$(1/2)mv^2$ U=mgy h=height

K+U=C mgh=mgy+$(1/2)mv^2$ or $(1/2)m(v\_0)^2$=U+K where v_0=initial velocity

Work=W=Fx and U=-W

Thus work is the distance traveled or moved by the component of the force in that direction, and potential energy is the negative of the work. Use the definition for work and the chain rule for derivatives to show that kinetic energy (energy of motion) is as given above. The chain rule is:

dv/dt=(dv/dx)(dx/dt)

A ball rolling on an incline will stay in motion until it attains the same height on another incline facing the first, even if the inclinations of the two inclines are not the same. If there is no second incline, the ball will never attain the original height and will therefore continue to roll forever, unless otherwise acted on by a force, like friction. For every force there is an equal but opposite reaction.

Notice that:

mgh=$(1/2)m(v\_0)^2$

Part 26

Let the closest star, Rigel Kentaurus, also called alpha centauri, be a metaphor for the earth, in that it is the closest star to the sun, and the third brightest in the sky and the earth is the only planet brimming with life and is the third planet from the sun. Thus we now consider the largest planet in the solar system (Jupiter) and this takes us to the brightest star in the sky, Sirius, alpha canes major, it is the fifth nearest star and Juptiter is the fifth planet. But let us associate with Jupiter as well Vega, it is the fifth brightest star. Sirius is 8.7 light years distant, and Rigel Kentaurus is 4.34 light years distant. 8.7/4.34=2.00…

If "the surface of the earth is the shore of the cosmic ocean" as Carl Sagan said, then the constellation Bootes carries its importance in the fact that it is "the boat-man". The brightest star in that constellation is Arcturus, which happens to be the fourth brightest star in our galaxy. The fourth planet is mars, and as it so happens, it is the only planet in our solar system we can colonize, in that mercury is so close to the sun that it is far too hot, venus has the same problem, but mars is the next planet after the earth, and the rest being gas giants, you can't really set foot on them. That is, we say that mercury, venus, the earth and mars are the terrestrial planets.

Part 27

First, here is a quote from the introduction of The Persian Wars, by Herodotus, by Francis R.B. Godolphin:

"A chemical formula is intelligible to anyone who knows the language; an isolated historical fact may correspond to a chemical symbol, but an historian's formula requires a great deal more than the juxtaposition of several such symbols to be intelligible."

There exists in Oregon a triad of species, as I like to call it, and they are:

Red Back Vole: by depositing its feces at the base of trees this small mammal innoculates the tree against disease allowing old growth forests to exist, becuase it eats a fungus that grows there.

Flying Squirrel: a flying mammal, it is very rare.

Northern Spotted Owl: At the top of the food chain of an old growth forest, it is an indicator species, or barometer for how the ecosystem is doing. It depends on old growth forests because it makes its nest in snags, or trees so old that the top has broken off and provides a flat surface for it to do so. This bird is endangered.

Part 28

The best way to explain this is to tell it in story form. I was looking for where standard temperature and pressure (STP) occurs the most, where STP is the temperature at which water freezes (32 degrees F) and the atmospheric pressure at sea level (one atmosphere). I knew California's coast was too warm and Washington state's too cold, as I have spent time in both, so I looked towards Oregon. Tillamook—where they make the fabulous cheese—was two degrees too warm on the average in January, so I estimated further north, somewhere between Nehalem and Astoria on the northern Oregon coast. The record high, by the way, for Oregon, was 119 degrees F in Pendelton on Aug 10 1898 and the record low was minus 54 degrees on Feb 10, 1933 in Seneca. The average of these two is 32.5 degrees F, just a half of a degree over freezing.

Now at STP the molar volume of a gas is 22.4 liters and one mole is 12 grams of carbon. What then, is the molar volume in quarts? Converting, you will find it is 23.67864693 qt for there are 946 mL in a qt and 1000 mL in a liter. Notice that 23.67864693 has all of the numbers between 3 and 9 except for five, after the decimal, and all of the numbers between 2 and 9 except for 5 in the whole number. It is equal to 11200/473 and the decimal part is 321/473. Now the question I ask is what is a quart? Well it is one fourth of a gallon. Then, what is a gallon? I looked in the dictionary and it said "origin unknown". Let us consider the molar volume then, 23.6 qt, and find the element whose density in pounds per quart equals the molar volume in quarts, and we find it is lead, chemical symbol Pb for plumbus, and hence the word plumbing because the Romans used lead for their plumbing. The density of lead is 11.4 g/ml, and a pound (lb) is 454 grams, and there are 946 ml in a quart.

Now in light of this a liter is one kilogram of water, at 4 degrees C, and a liter is a cube with sides one tenth of a meter, and a meter is one ten millionth of the distance from the north pole to the equator. Thus, it would seem lead and water are related in more way than one. And what are Nehalem and Astoria, these cities in Oregon? They bring to mind Akkadia and the Nefilim of Nibiru, the latter being held by Sumerian myth to be the people and planet responsible for the creation of our planet and ourselves respectively as translated from cuniform tablets.

Part 29

Project Genesis defined: It is Assyrian in nature. Here is a quote from Carl Sagan's Cosmos.

"In ancient times, in everyday speech and custom, the most mundane happenings were connected with the grandest cosmic events. A charming example is an incantation against the worm which the Assyrians of 1000 B.C. imagined to cause toothaches. It begins with the origin of the universe and ends with a cure for the toothache."

Now the elements are in no way mundane, but as we began connecting their properties to things on a "cosmic scale" we perhaps shall soon see that everyday events are connected to those connections. We are operating on a grander scale at this point in time, than the mundane, like the constructions of pyramids, and the origins of agriculture, but soon we will hopefully understand what brushing our teeth in the morning means, in a deep sense and in relation to the structure of the universe, and as such unlock the mysteries of origin and destiny. Our advantage is that we are operating empirically as opposed to our Assyrian friends. We are doing much more than searching for a cure for the toothache, but are trying to save humanity, by opening the key to reproducing the system that keeps us alive.

Part 30

We are all familiar with the pyramids in Mexico

There is Teotihuacan in Mexico City and

Giza outside of Cairo, I believe

I have been told that they built pyramids in India

and that all of these lie in the so called "sun belt"

Now looking at a map, Teotihaucan and Giza are both within the

tropic of cancer and the first lattitude line after that, plus 30 north. I have not
been able to find anything about the pyramids of India on the internet.
Now the lattitude plus 30 degrees is interesting because 30 degrees is the angle
formed by drawing in the altitude of an equilateral triangle. Cairo is exactly on
this lattitude, and, as I said, so are its pyramids.

Until about 8,000 B.C.
Man followed the herds as he hunted
which followed the seasons
Some ten thousand years ago he began
to settle in communities and farm and ranch
and began to work regularly
This most likely began in the middle east, jericho being
the oldest known site of a cultivating community
Its latitude is about the same as central California's, the most productive farming
place in the world today
And as early a 1500 B.C. in the middle east, men began to smelt iron ore. This
represented a revolution in weapon and tool making. Iron is the second most

abundant metal in the earth's crust comprising 4.7% of it. Heavy it holds a sharp edge. The earth's crust is mostly silicon, used to make integrated circuits, diodes, and transistors. The most abundant metal is aluminum at 7.5% of the earth's crust. As can be seen, all we need is there, and silicon is mostly made useful in silicon valley of california. There it is doped with boron or phosphorous. Computer chips are made of silicon. Thus Iron is useful for agriculture and silicon for technology. An atom of iron is twice as heavy as an atom of silicon. Silicon is 25.7% of the earth's crust.

When they speak of Atlantis, perhaps it is a metaphor for Ebla which was on about the same parallel as Los Angeles, which seems to have had a parallel function, as their economy was based on the manufacture of textiles and metal works.

Of importance, I believe, is what was going on when Vega was the "pole star", Alpha Lyra, the fifth brightest star in the sky. Polaris, today's pole star is relatively faint. Vega will again be the pole star in 14,000 A.D. Vega was the pole star some 12,000 years ago, when the "land bridge" existed during the last ice age when the American Indians theoretically crossed it to the New World. It would seem to me that Vega and Polaris switch the honorable position every 13,000 years as they are in opposition in the precession of the Earth's axis, which completes a full cycle every 26,000 years. Alpha Draconis, or Thuban, the brightest star in Draco was the pole star some 4600 years ago when the Egyptians were building their pyramids.

A.D. 100's Ptolemy offers the idea of epicycles to explain the retrograde motion of the planets.

About 310B.C.-230B.C. Aristarchus determines the distance to the moon and that the sun is very far away using parallax.

(276-195?B.C) Eratosthenes determines the circumference of the earth by the angle cast by the shadow of a stick and the absence of a shadow in a well in another location.

1473-1543 Copernicus puts forward his model of a solar system where the earth is not at the center of the universe, but that goes around the sun along with the other planets. The model explains the retrograde motion of the planets.

1564-1642 Galileo discovers four natural satellites orbiting Jupiter, and thus verifies the idea of Copernicus that the Earth is not at the center of the universe.

1571-1630 Based on the observations of Tycho Brahe, Kepler formulates the laws of planetary motion.

1642-1727 Newton makes his universal law of gravitation from which Kepler's laws can be derived. He invents differential calculus and integral calculus, the latter simultaneously with Liebnitz.

Part 31

Logica
Logica System 0

There can be no so such a thing as position in infinitely extended space, for if there are only (A,B,C) then A is referenced with respect to B and B with respect C and C cannot be referenced with respect to anything independent from A or B, therefore A has no position. If there is (A,B,C…) then A is referenced by B, and B by C, and C by D, and so on forever. A is never referenced. Another way of looking at this is a circle is the collection of points equally distant from some point called the center. It follows that in infinitely extended space the center is everywhere. And space must be infinite in extension because there always has to be another side to anything, even if it is nothing. Change in position with respect to time is motion. Anything in motion cannot exist for any amount of time at any position, otherwise it has stopped. Yet we say what the position of something is after a certain amount of time given its speed. Thus motion is impossible, even though we see it happen. But that is O.K. because we have shown there can be no such a thing as position.

Logica System 1

Prove: God is artificial.

Postulate 1: For ourselves, the planets, the stars (the galaxies are made of stars), and light to be here, something had to come into existence from nothing, uncaused, yet that is impossible. The only other alternative is that something has always been here, and that does not make any sense. In this sense nothing can exist.

Axiom 1: It cannot be proved that what we see, hear, smell, feel or taste exists.

Definition 1: God—the source of existence.

Theorem 1: Man made God (by postulate 1, axiom 1 and, definition 1).

Definition 2: That which is man made is artificial.

Theorem 2: God is artificial (by definition 2 and theorem 1). QED.

Logica System 2

Prove: Nature is God.

Axiom 1: Objects require their opposite to be defined. i.e. matter is defined in terms of space: it is that which occupies space and has mass.

Theorem 1: The set of everything is beyond explanation because there exists nothing outside of it to define it (by axiom 1).

Definition 1: God—that which is beyond explanation.

Definition 2: Nature—everything.

Theorem 2: Nature is beyond explanation (by theorem 1 and definition 2).

Theorem 3: Nature is God (by theorem 2 and definition 1). QED.
Hypothesis: For every thesis the antithesis exists, even if neither make sense. In other words, to quote Nielhs Bohr: "The great truth is a statement whose opposite is also true." (By induction contrasting the last theorems of logica system 1, and logica system 2).

The truths that we accept, are, more than anything, dependent on our definitions. Different cultures represent different logic systems. Logica System 0, Logica System 1, and Logica System 2 represent different hypothetical cultures. Notice how one and two come to a seemingly opposite truth. Logica System 0 we will call universal: its truths are accepted by all cultures, let us say.

Logica System 3

In order to define A we might say it is BC. But to define BC, we must say that B is, for example, DE and C is EF. We then must define DE and EF, and the process can go on forever, so we might as well be content that A is A, and leave it at that.

Logica System 4

Things only have the meaning we give them. For example, if I say lets discuss fringe philosophy, science, and technology, one person might interpret that as let us discuss fringe philosophy, fringe science, and fringe technology, and another as let us discuss fringe philosophy, with science and technology not on the fringe.

Logica System 5

All is one. Even a blank page is a drawing of nothing. In art emptiness is just as important as lines. In music silence is just as important as sound.

Logica System 6

Since there is subject and object, observer and the observed, all contemplation results in paradox, for there is no way to determine whether or not anything we see actually exists.

Life Path One: Keep your mind always open to thesis and antithesis until the synthesis is encountered. Thesis is one base corner of a triangle and antithesis is the other base corner of the triangle, and the synthesis is the apex of the triangle. Each synthesis becomes a new premise for which there is another antithesis. With the synthesis of these, we have a new triangle whose base corner is the apex of the previous triangle. As such thought is complexes of triangles. Eventually old contradictions resolve.

Life Path 2: Believe in a thesis by arguing its validity. Reject the antithesis. Continue to develop the thesis, as more are born that you support.

Logica System 7

Since every statement requires a statement to prove it, one has to start with something that cannot be proved, in order to prove something. In this sense, nothing can be proved. It follows that, in order to prove something we must either:

a) have faith in something.

b) have the belief that the truth is self-evident.

Logica System 8

The statement: "nothing can be proved" is self contradictory, therefore its negation must be true: "anything can be proved".

One could say that a bee's honeycomb is natural. In this sense a human's house is natural. One could say that a human's house is technology. In this sense a bees honeycomb is technology. Or one could say that a bee's honeycomb is natural and a human's house is technology. All of these statements are true from a different point of view. The fourth permutation is that a human's house is natural and a bee's honeycomb is technology.

Part 32

The sun radiates 1.2E34J/yr and

plancks constant is 6.626E-34(J)(s) (minimum energy packet)

(1.2E34J/yr)(6.626E-34J*s)(yr/3.15E7s)=2.5E-7J^2

(1.01g H/mol)(mol H)/(6.02E23 atoms H)=1.68E-24gH/atom

rest energy of a hydrogen atom:

(1.68E-24g)(3.0E8m/s)^2(kg/1000g)=1.5E-10J

(2.5E-7J^2)/1.5E-10J=1.67E3J

The specific heat of water is 1.00cal/g*deg C and

1.67E3J=3.99E2cal

the normal human body temperature is 37 deg C

(3.99E2cal)(deg C)(g)/(1.00cal)(37deg C)=10.78gH_2O

10.78g=1.078E-2Kg

(1.078E-2)(15807m)(9.8m/s^2)=1.67E3J

15807m=16km

which is right in the middle of that narrow layer of atmosphere between the troposphere and stratosphere, called the tropopause.

Now 16km is 19 kHz, or very low frequency (VLF) which corresponds to wavelengths between 100km-10km. It is a radio frequency and happens to be the frequency at which radio waves can penetrate water 10-40 meters deep (10-20kHz) and thereby is used by submarines near surface for communication. Now this was based on the speed of light in a vacuum, but in anycase we have connected human activity to natural law, physical and biological, and are thus bordering on psychohistory.

Now the normal human body temperature is about the difference between the coldest day and warmest day of the year in Red Bluff, California, the agricultural hub of America, if not the world. That is the average temperatures swing between about freezing (of water), 0 degrees C and 37 degrees hot throughout the course of a year. In fact between 1961 and 1990 the average yearly normalized temperature was 17.1 degrees C. (37 deg C–0 deg C)/2=37/2=18.5degC, an accuracy within 92%. Red Bluff is in Tehama County, which is at about 40degN, 122degW. (See www.worldclimate.com)

Part 33

We apply the triangulation of Chris Darrow, where three relevant locations are chosen on a map, are connected and the associated power spots pertaining to the corners are revealed. Bear in mind that a triangle is the minimal structure that encloses an area.

I have chosen Nehalem, Oregon as it is where I estimate that standard temperature and pressure occurs most frequently, that is freezing temperature of water and 1 atmosphere of pressure. I have chosen as a second point Red Bluff, California as its annual average temperature is the mean of freezing and the normal body temperature, 37 deg C. I have chosen as the third point Bend, Oregon, as it is "land's end" in the high desert the last city before the vast expanse to Idaho, nearly. Connecting these points we have a triangle who's "center of gravity" seems to be the Cascade Mountain range Summit at Crescent Lake: http://www.crescentlakeresort.com/

So far we have been very accurate where the solar system is concerned, but we just don't have accurate data to do anything on a scale as grand as the galaxies, their clusters and the dimensions of the universe, so I am left with no other alternative than to be silly in this area.

So, let me be an artist/comedian for just a minute and give the intuition free reign to draw a fuzzy picture in poem form:

The hubble constant is no doubt somewhere between 50 and 100km/sec/Mpc, but is more than likely 60 to 80km/sec/Mpc, and no doubt 70km/sec/Mpc +/- 5%. It is the expansion rate of the universe. The dinosaurs became extinct about 65 million years ago. Now let us calculate how much the universe has expanded within that amount of time at that rate.

The time: 65E6yrs

Expansion rate: 70km/sec/Mpc

There are 3.15E7sec/yr

The speed of light is: 3.0E10cm/s=3.0E5km/s

Km/ly=9.45E12

Pc=3.26ly

Hubble constant (H_0)=7.0E(-11)(ly)/yr/ly

Theoretical age of the universe by calculating it based on the temperature to which it has cooled (2.76 degrees K): 15 billion yrs. That makes it 15 billion ly in radius.

We have that:

v=D(H_0)=15E9(ly)(7.0E(-11)ly/yr/ly=1.05(ly)/(yr)

Where d is distance and v is velocity, the expansion speed of the universe.

Experimental age of the universe by multiplying the speed of light by the furthest seen objects, which are 10 billion light years away: 10 billion years. That gives us:

V=D(H_0)=10E9(ly)(7.0E(-11)ly/yr/ly=0.7(ly)/(yr)

Thus by the former the universe has expanded 1.05(65E6)=6.8E7(ly)

And by the latter: 0.7(65E6)=4.6E7(ly)

The average of these two is 5.7E7(ly)

which is on the order of the Pegasus I cluster of galaxies in diameter.

When the dinosaurs went extinct, the Virgo cluster (core of the local super cluster of galaxies, and nearest cluster of galaxies) was just emitting the light that we are receiving today as it is 64 million light years away from us, it contains the stunning sombrero galaxy.

The distance the Bootes galaxy cluster has receded since the dinosaurs went extinct:

(39400km/s)(65E6yrs)=(1.3E(-1)(ly)/(yr))65E6yr=8.45E6(ly),

and how far the Hydra galaxy cluster has receded since the dinosaurs went extinct:

(60,600km/s)(6.5E6yrs)=(2.0E(-1)(ly)/(yr))(65E6yr=1.3E7(ly)

which is the approximate diameter of the Virgo cluster, and Cancer cluster, Leo cluster and Gemini clusters of galaxies.

Why be so interested in the correspondences of the Virgo cluster's various parameters with the extinction of the dinosaurs? Because it would seem the dinosaurs sudden disappearance gave rise to the intelligent mammalian life we are today, and as such, with the same laws working throughout the universe, odds may be that intelligent life elsewhere would have descended from something reptilian not mammalian, as the extinction of the dinosaurs was a fluke, like cause by the impact on earth of an asteroid. Thus the Virgo cluster may correspond to the success of reptilian life where they did not have a chance to evolve here.

Part 34

Earth-moon distance: 3.84E10cm=R
Earth-sun distance: 1.496E13cm=r
Earth radius: 6.38E8cm=E_r
Moon radius: 1.738E8cm=M_r

Apparent diameter moon=apparent diameter sun=0.5 deg

(360)/(0.5) =720 moon diameters/celestial equator = 720 sun diameters/celestial equator

$(E\_r)/(M\_r)=11/3$ r/R=4675/12

(4675/12)(11/3)/720=2

$(E\_r/M\_r)(r/R)(D/C)=(O\_2)/(CH\_4)=(r\_u)/(r\_s)$

C=degrees in circumference of a circle
D=apparent diameter of moon=apparent diameter of sun in degrees.

…r_u=mean distance from sun of Uranus
…r_s=mean distance from sun of Saturn

We have essentially shown that:

$((E\_r)/(M\_r))((1$ deg in rad))=(8(pi rad))(R/r)

Where, again, E_r=earth radius, M_r=moon radius
C=radians in circumference of a circle, R=earth-moon distance
r=earth-sun distance. The last means that

$((E\_r)(r))/((M\_r)(R)(360))=4$

Now 360 are the degrees in a circle and
4 weeks is a complete revolution of the moon
about the earth more or less on the average
(time between new moons). 360 degrees are
convenient for the units in which to divide a circle,
because of its divisible properties. (i.e. it is divisible by
120, 60, 45, 30, and 90, into whole numbers which are the angles in special triangles.)

Part 35

The specific heat of water, the energy required to raise the temperature of a gram of it one degree centigrade is (1.00 calories/gram-degree centigrade).

The luminosity of the sun is 3.826E26J/s.

(365days)(24hr./day)(60min/hr.)(60s/min)=3.15E7 seconds per year.

The annual output of energy by the sun in light, that is, over one complete revolution of the earth about the sun is:

(3.826E26J/s)(3.15E7s)=1.2E34J/yr

One calorie (1 cal)=4.184 Joules (J)

(1.2E34J)(cal/4.184J)=2.87E33cal

That is the annual solar output of energy in light in calories.

Water freezes at zero degrees C at 1 atmosphere of pressure, or, at sea level in other words, and the normal human body temperature is 37 degrees C. Thus using our specific heat of water as a unit factor, we have, since 37C-0C=37C:

(2.87E33cal)(1/37C)(g-C)/cal)=(7.76E31g)(Kg/1000g)=7.76E28Kg

A so-called enthalpy calculation will show that a gram of sugar as made by plants (C6H12O6) will produce 1.56E4J of energy when burned in oxygen. Thus,

(7.76E28Kg)(1.56E4J/g)(1000g/Kg)=1.2E36J

Thus if over the complete revolution of the earth about the sun, the total energy output of the sun in light, raises water from freezing to the normal human body temperature at the freezing of water and sea level, then that would correspond to 7.76E28Kg of water, which in mass of photosynthetically produced sugar, burned, is 1.2E36J of energy.

Now Jupiter has a mass of 1.9E27Kg, and a mean orbital velocity of 13060m/s. This means its kinetic energy due to its orbital motion, or energy of motion, as given by $(1/2)mv^2$ is:

1.62E35J

$(1.2E36J)/(1.62E35J)=7.41$

$7.41\sim Au/Al=196.97/26.98=7.3$

Where Au is the molar mass of gold, and Al is the molar mass of aluminum. As there are no isotopes of gold or aluminum, this ratio is independent of the sample. Gold is about 7 times as dense as aluminum at earth temperatures and pressures.

Gold is special for its unique electrical properties and aluminum is the most abundant metal in the earth crust.

Absolute zero, or as cold as it gets in other words, is—273 degrees C, and, the normal human body temperature is, as we said, 37 degrees C. That makes us 8.378 times warmer than absolute zero, which corresponds to the ratio of platinum to sodium: $195.08/22.99=8.4854=Pt/Na$. I believe this relationship to be the key to maintaining a habitable habitat (temperature) for humans inside, and perhaps even outside the ship. (Consider platinum and sodium).

Part 36

$(C6H12O6)/(H2O)=10$

(density of gold)/(density of aluminum)$=7$

$(10+7)/2=8.5\sim Pt/Na=\ldots$

The kinetic energy of the Earth, K_e is
$K\_e=(1/2)(5.976E24kg)(29790m/s)^2=2.65E33J$
and that of Saturn is K_s:
$K\_s=(1/2)(5.68598E26)(9640m/s)^2=2.64E34J$

$(K\_s)/(K\_e)=(2.64E34J)/(2.65E33J)=9.96\sim10=C6H12O6/H20$

glucose compared to water.

Part 37

The golden ratio is the ratio such that the whole portion to the larger portion equals the larger portion to the lesser, or in other words, we have that a/b=b/c when a=b+c.

$(a/b) = (b/c)$ when $a = b + c$

$(a/b) = (b/c)$

$ac = b^2$

$a = b + c$

$c = a-b$

$a(a-b) = b^2$

$a^2-ab = b^2$

$a^2-ab-b^2 = 0$

$(a^2/b^2)-(a/b)-1 = 0$

the last equation is a quadratic in a/b.

$(a^2/b^2)-(a/b) = 1$

$(a^2/b^2)-(a/b) + (1/4) = (5/4)$

$$((a/b)-(1/2))^2 = (5/4)$$

$$(a/b)-(1/2) = (sqrt(5))/2$$

$$a/b = (sqrt(5) + 1)/2$$

978-0-595-36540-1
0-595-36540-X

www.ingramcontent.com/pod-product-compliance
Lightning Source LLC
Chambersburg PA
CBHW021545200526
45163CB00015B/1785